THE AIRPORT BUILDERS

BY
JAMES E. KELLY
AND
WILLIAM R. PARK

DRAWINGS BY
JOEL SNYDER

▲ ADDISON-WESLEY

THE AIRPORT BUILDERS

Addisonian Press Titles
by James E. Kelly and William R. Park
THE ROADBUILDERS
THE AIRPORT BUILDERS

An Addisonian Press Book

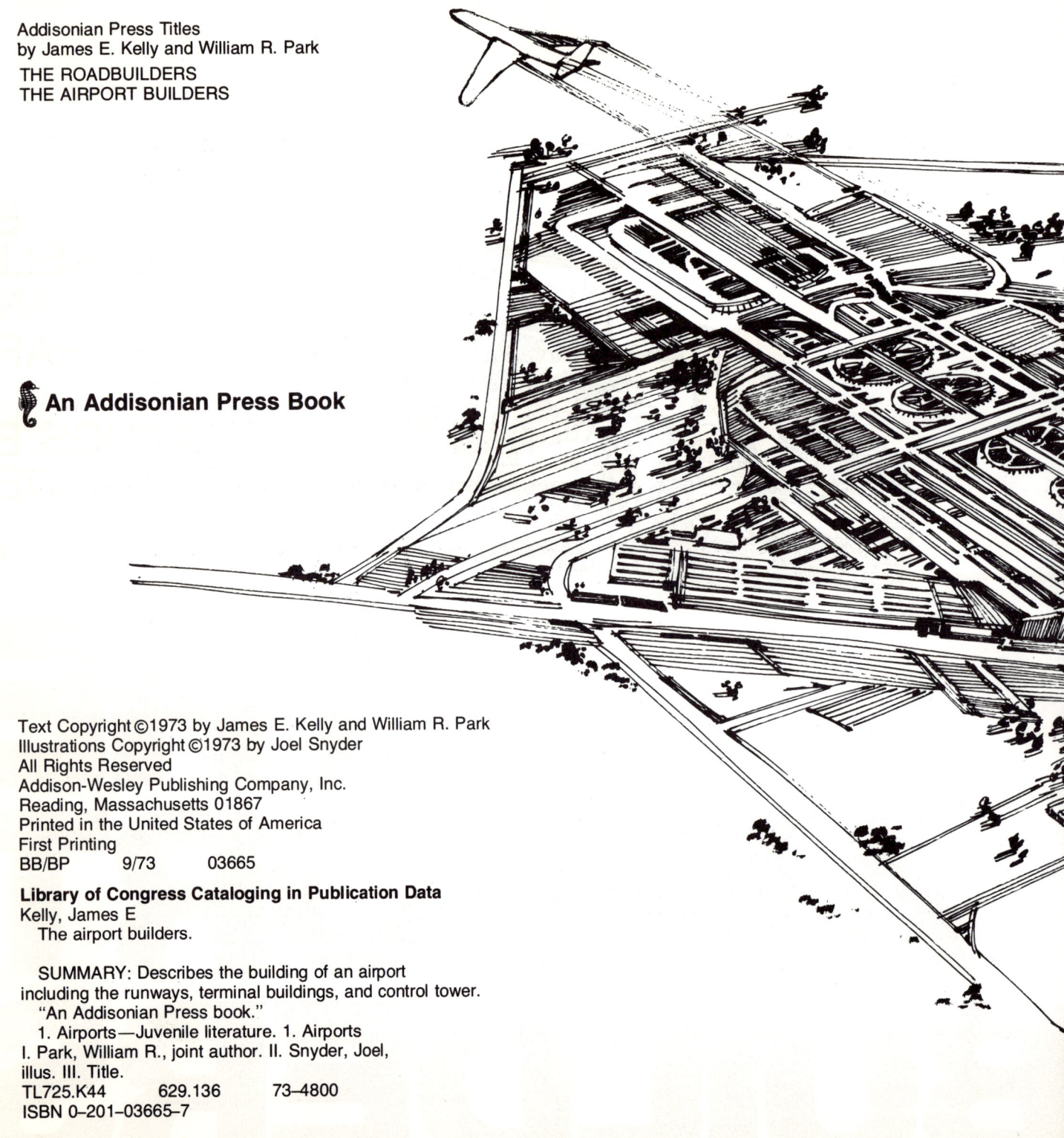

Text Copyright ©1973 by James E. Kelly and William R. Park
Illustrations Copyright ©1973 by Joel Snyder
All Rights Reserved
Addison-Wesley Publishing Company, Inc.
Reading, Massachusetts 01867
Printed in the United States of America
First Printing
BB/BP   9/73   03665

Library of Congress Cataloging in Publication Data
Kelly, James E
  The airport builders.

  SUMMARY: Describes the building of an airport
including the runways, terminal buildings, and control tower.
  "An Addisonian Press book."
  1. Airports—Juvenile literature. 1. Airports
I. Park, William R., joint author. II. Snyder, Joel,
illus. III. Title.
TL725.K44   629.136   73–4800
ISBN 0-201-03665-7

AIRPORTS ARE BUSY PLACES!

Airport builders must plan for many things. Every day many people will come to the airport. And lots of people will work at the airport.

The builders must plan buildings for all these people and for the airplanes.

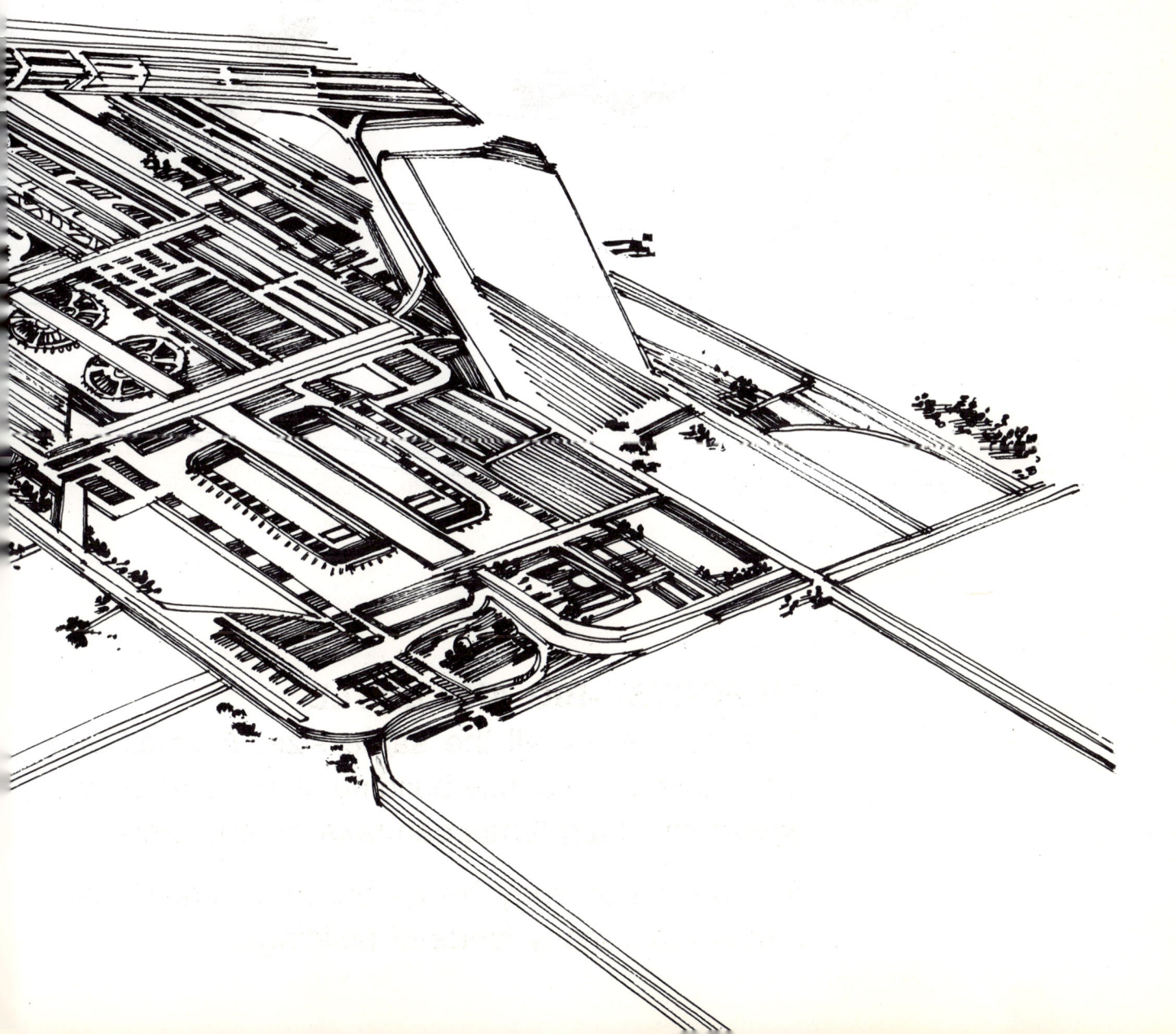

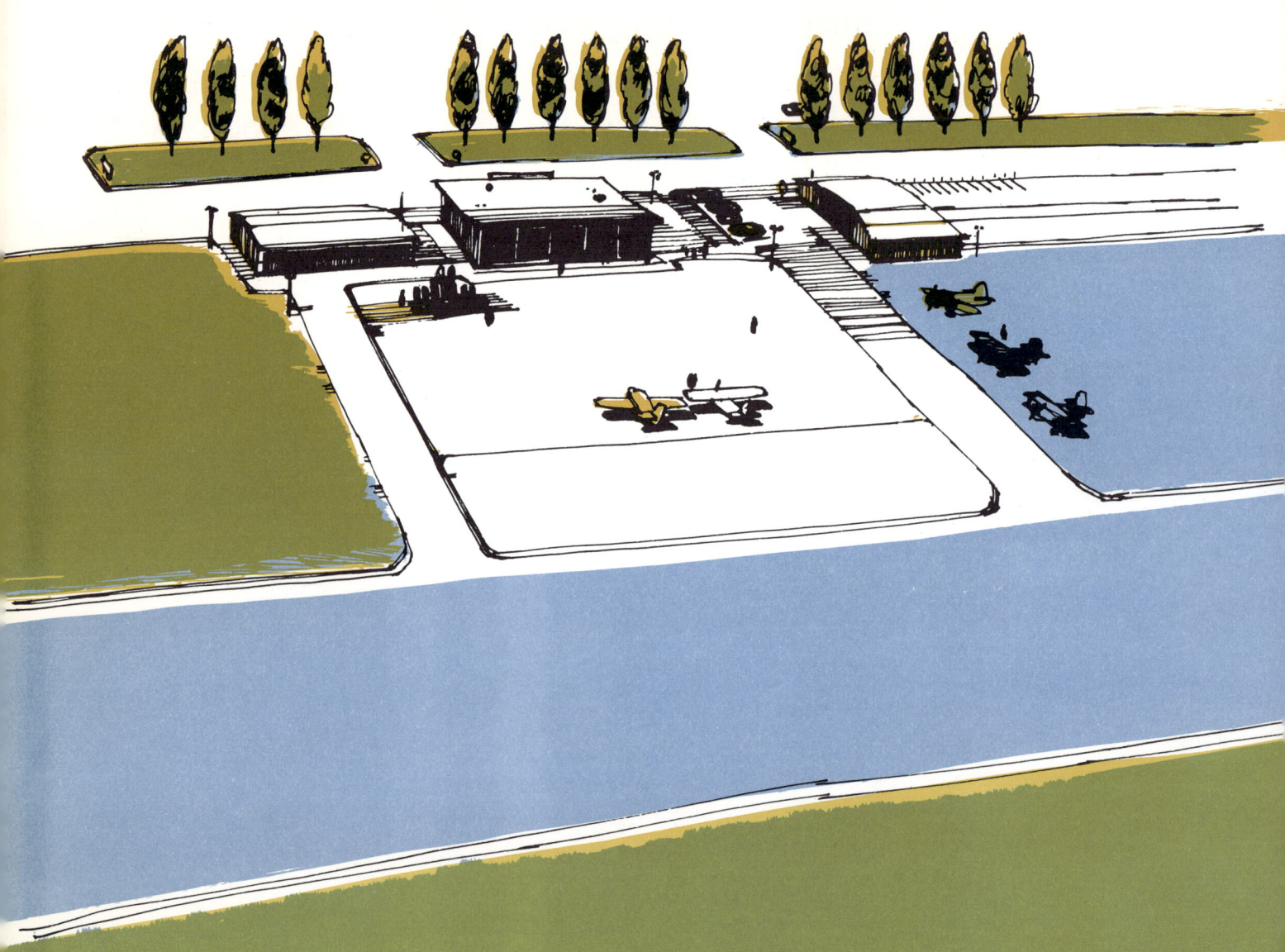

AIRPORTS—BIG AND LITTLE

Airports are not all the same size. A small one may have just a single building. It will have a field where small airplanes can take off and land.

A large airport may be as big as a small town. It may have many kinds of buildings.

AIRPORT RUNWAYS

Airplanes take off from and land on *runways*. Runways look like long, straight roads. They must be flat and smooth because airplanes travel very fast on them.

A small airport may have only one runway. At large airports there may be many runways.

AIRPORT MODELS

Before the airport is built, the builders often make a *model* of the airport. Then the builders can see if they have planned everything right.

Now the *plans* for the airport are drawn.

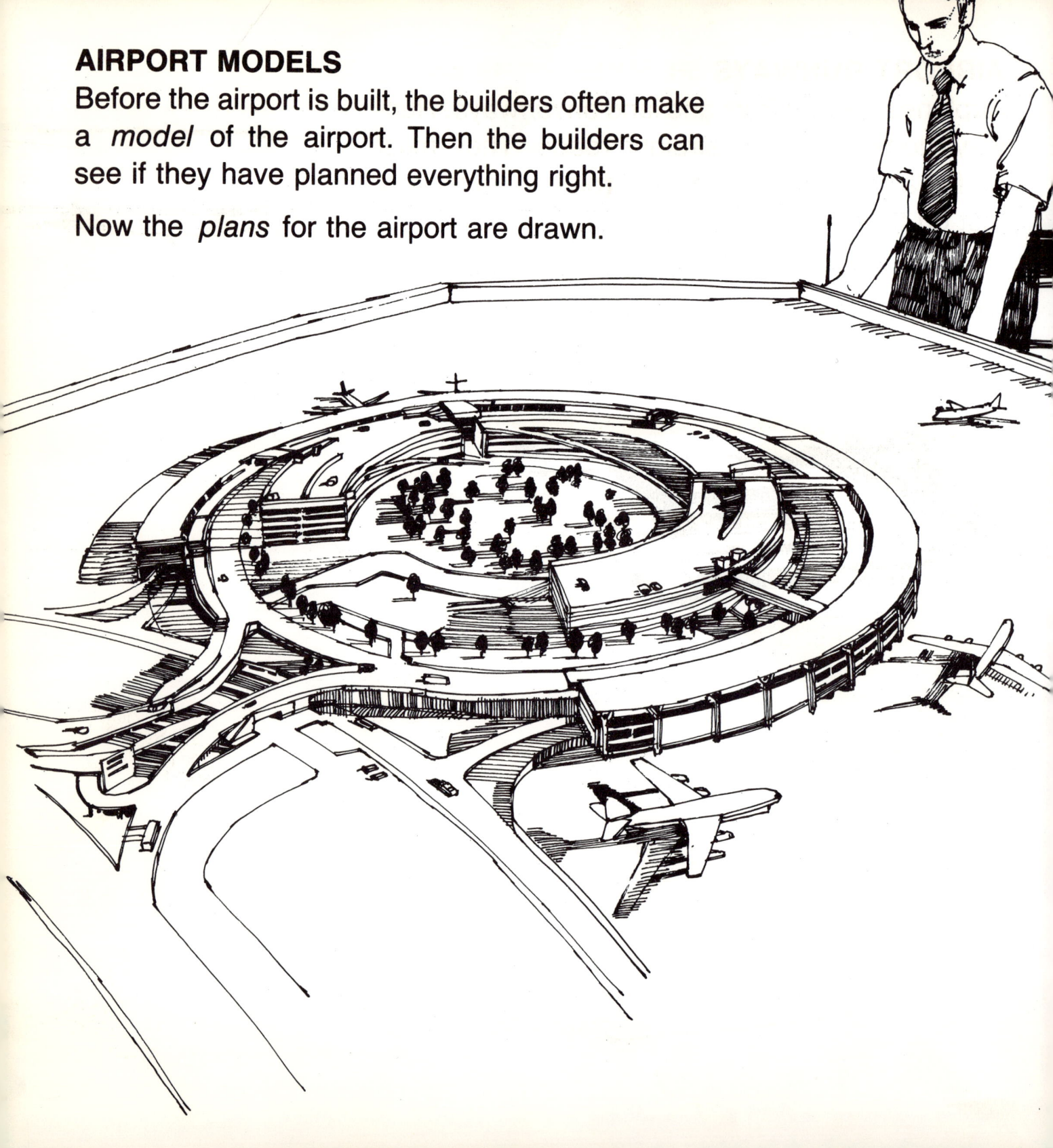

PLANNING THE AIRPORT

The plans are drawings of the buildings and runways. The sizes of the buildings and runways are marked on the drawings.

The people who prepare the plans for the buildings are called *architects*. The ones who prepare plans for the runways are called *engineers*.

When the plans are finished, it is time to clear the land for the runways.

CLEARING THE LAND FOR THE AIRPORT

Sometimes trees and buildings are in the way and have to be removed.

The engine of a powerful tractor roars as it pushes down a large tree. The tractor is called a *crawler tractor*. It moves along on tracks instead of regular wheels. There are big steel knives on the back of the tractor. They are called rippers. Before the tree is pushed down, the rippers are put in the ground and pulled around the tree. The rippers cut the roots and make the tree easier to push over.

Small buildings are also pushed down with big tractors. But big buildings must be knocked down with a *wrecking ball*. It is fastened onto a *crane*. Cranes look like machines with big fishing poles raised high in the air. But instead of fishing line, the cranes have strong rope made of steel wire. When the wrecking ball is finished, there is nothing left but a big pile of bricks and concrete.

Another kind of crawler tractor comes along. It is called a *tractor loader*. The tractor loader loads the bricks, concrete, and trees in trucks to haul them away.

THE CONTROL TOWER

There is one very tall building at the airport. It is the *control tower*.

The top of the control tower is called the *control cab*. It is a big room with glass walls on all four sides. The control tower is near the center of the airport.

The people who work in the control cab are called *air traffic controllers*. They are traffic policemen for airplanes. They direct the air traffic. The air traffic controllers talk to the airplane *pilots* by radio.

Pilots are the people who fly the airplanes. Air traffic controllers tell pilots when and where to take off and to land. They also tell pilots where to go when the planes are on the ground.

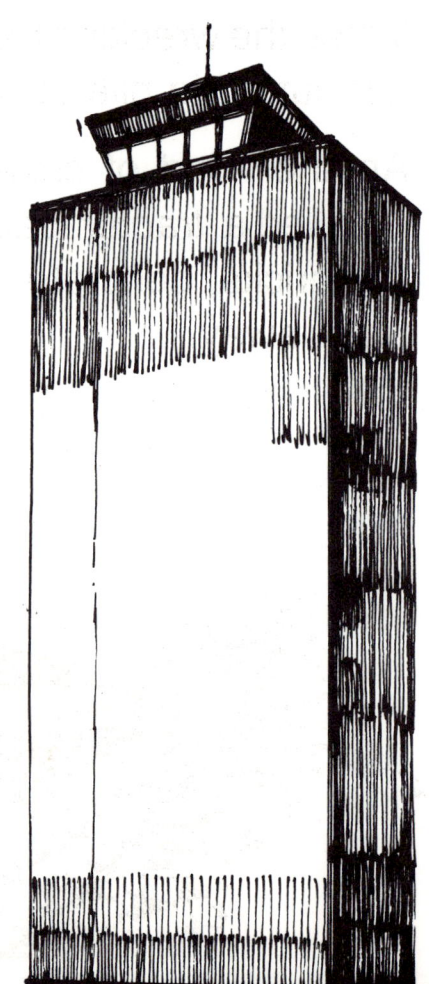

BUILDING THE CONTROL TOWER
Building the control tower begins with digging a deep hole for the *foundation*. *Front-end loaders* scoop big bucketsful of dirt out of the ground.

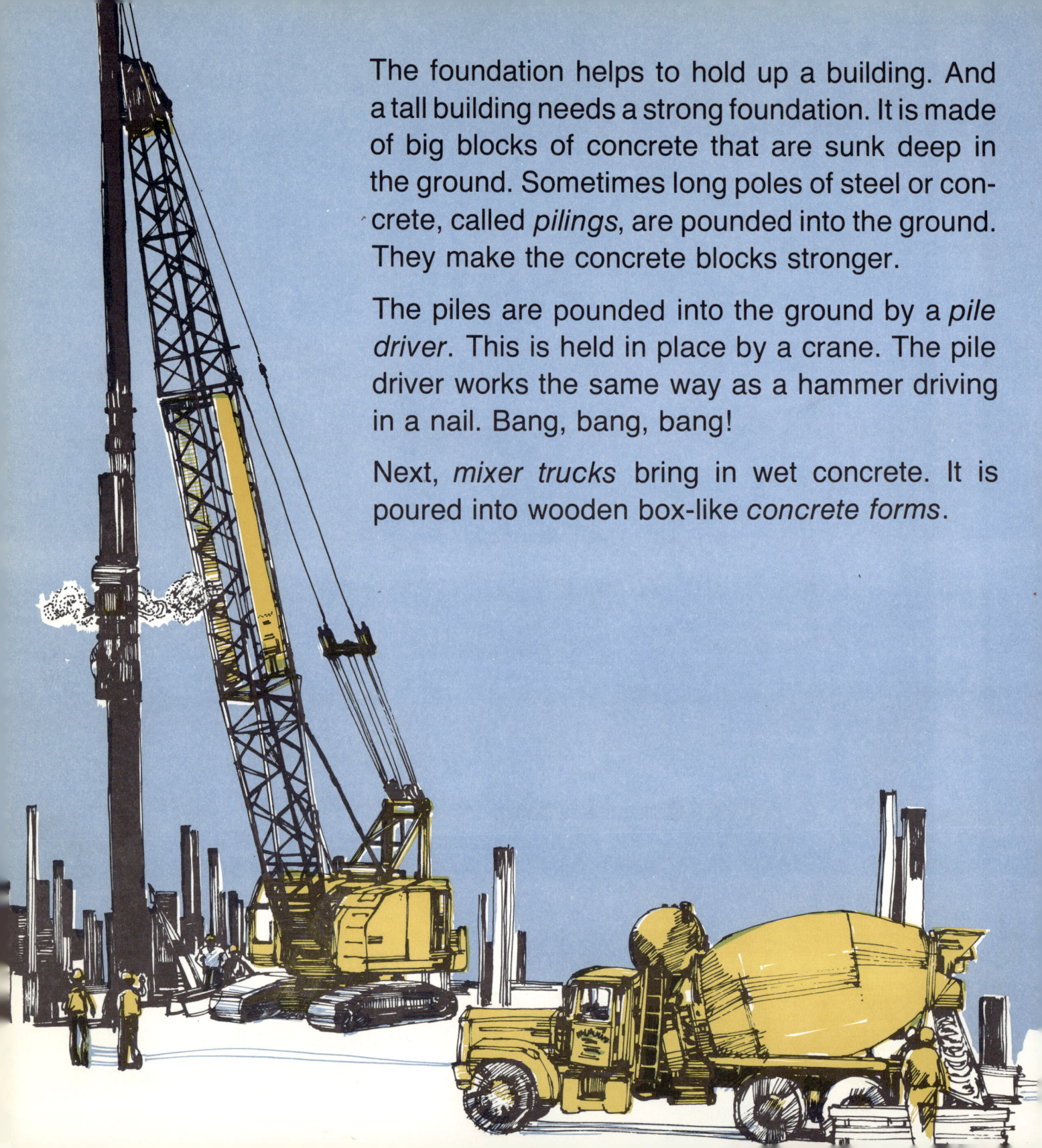

The foundation helps to hold up a building. And a tall building needs a strong foundation. It is made of big blocks of concrete that are sunk deep in the ground. Sometimes long poles of steel or concrete, called *pilings*, are pounded into the ground. They make the concrete blocks stronger.

The piles are pounded into the ground by a *pile driver*. This is held in place by a crane. The pile driver works the same way as a hammer driving in a nail. Bang, bang, bang!

Next, *mixer trucks* bring in wet concrete. It is poured into wooden box-like *concrete forms*.

The building will have many steel beams in it. Those that stick straight up are called *columns*. The cross pieces are called *girders*. All the steel beams are lifted into place by huge cranes. The crane lifts the beams high in the air. Workmen then fasten them together with steel bolts.

The framework of steel beams looks like a huge jungle gym. But it is so big that only giants could play on it.

Then the outside of the control tower is covered with panels made of concrete, or of steel and plastic. The cranes lift the panels into place.

The control cab is built in sections on the ground and raised to the top by the cranes. There workmen put it together.

THE AIRPORT TERMINAL

The main building at an airport is called the *terminal building*. In most airports this is where passengers will go to board their planes.

Terminal buildings are different sizes and shapes. But they all provide places for the people, airplanes, cars, and buses.

A concrete *apron* is on one side of the building. Here airplanes park next to a gate where the passengers can come and go.

Cars and buses drive on the other side of the terminal building.

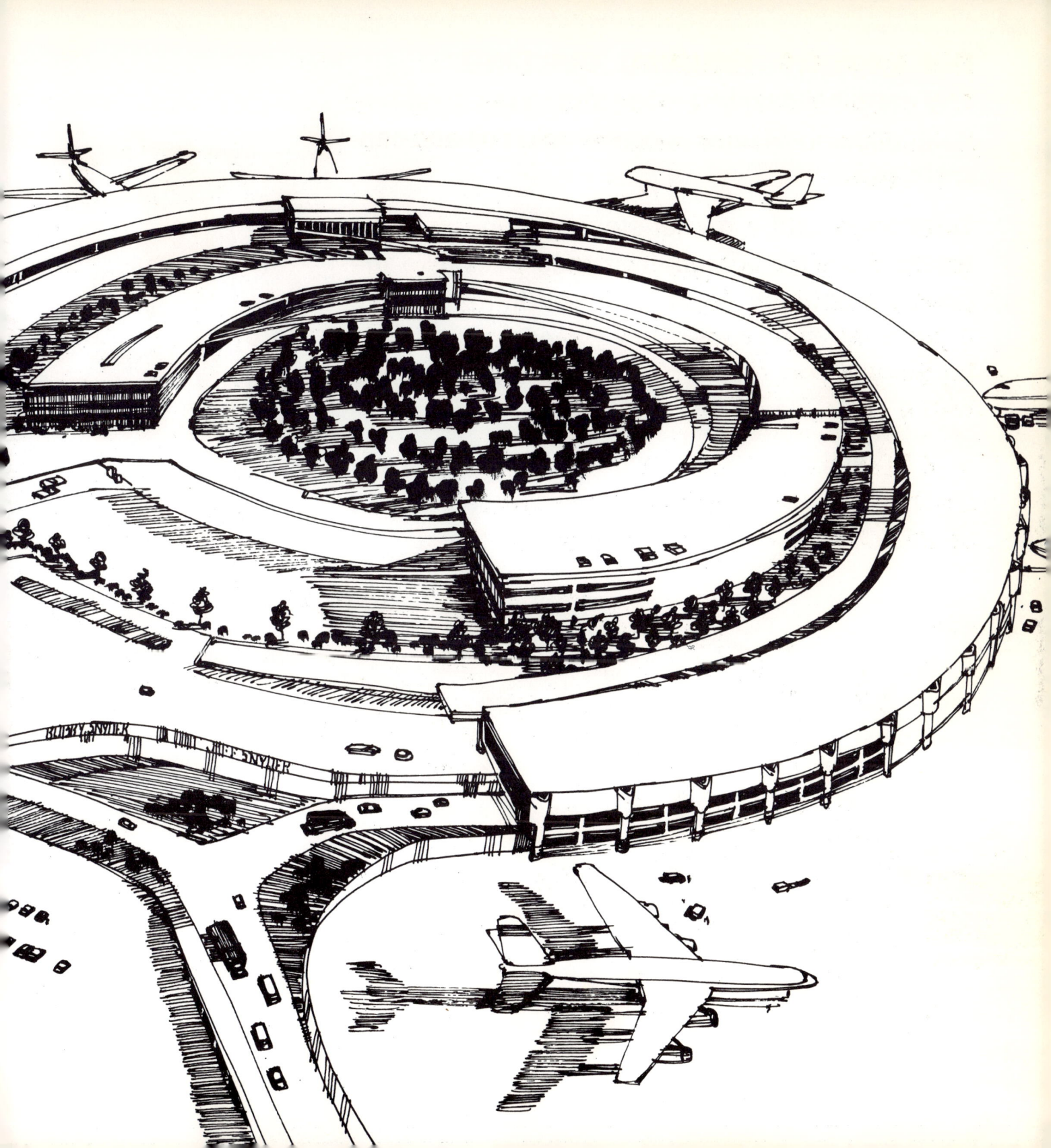

BUILDING THE TERMINAL BUILDING

The terminal building does not need a strong foundation. It will have concrete columns and concrete walls.

Ironworkers put a row of steel rods in the place where the walls will go. The rods will make the concrete stronger. Next, concrete forms are put around the steel rods. Now everything is ready for the wet concrete to be pumped into the forms with a *concrete pump*.

When dry, the concrete becomes hard. The forms are then removed. Later, concrete beams are lifted up and bolted across the concrete columns.

Holes are left in the concrete for the pipes and wires. These will run all through the building. But they will be hidden after the building is completed.

There must be electric wires for the lights, office machines, and electric signs.

Plumbers also put in pipes for water.

On top of the building, workmen spread a layer of hot, black asphalt on the roof with mops. Next they cover the roof with a layer of black felt, which is like the material used to make men's hats.

The workmen put on five layers of felt covered by five layers of asphalt before they are through. Then they cover it all with a layer of rock chips.

The roof will not leak now. And it will also keep out the heat and cold.

Inside the building, *carpenters* are busy sawing boards and nailing them together. The carpenters are building walls for offices and shops.

Painters are at work painting. And other workers are busy putting up signs. The signs will tell people where to find things like the right gate, where to buy their tickets and where to find their baggage.

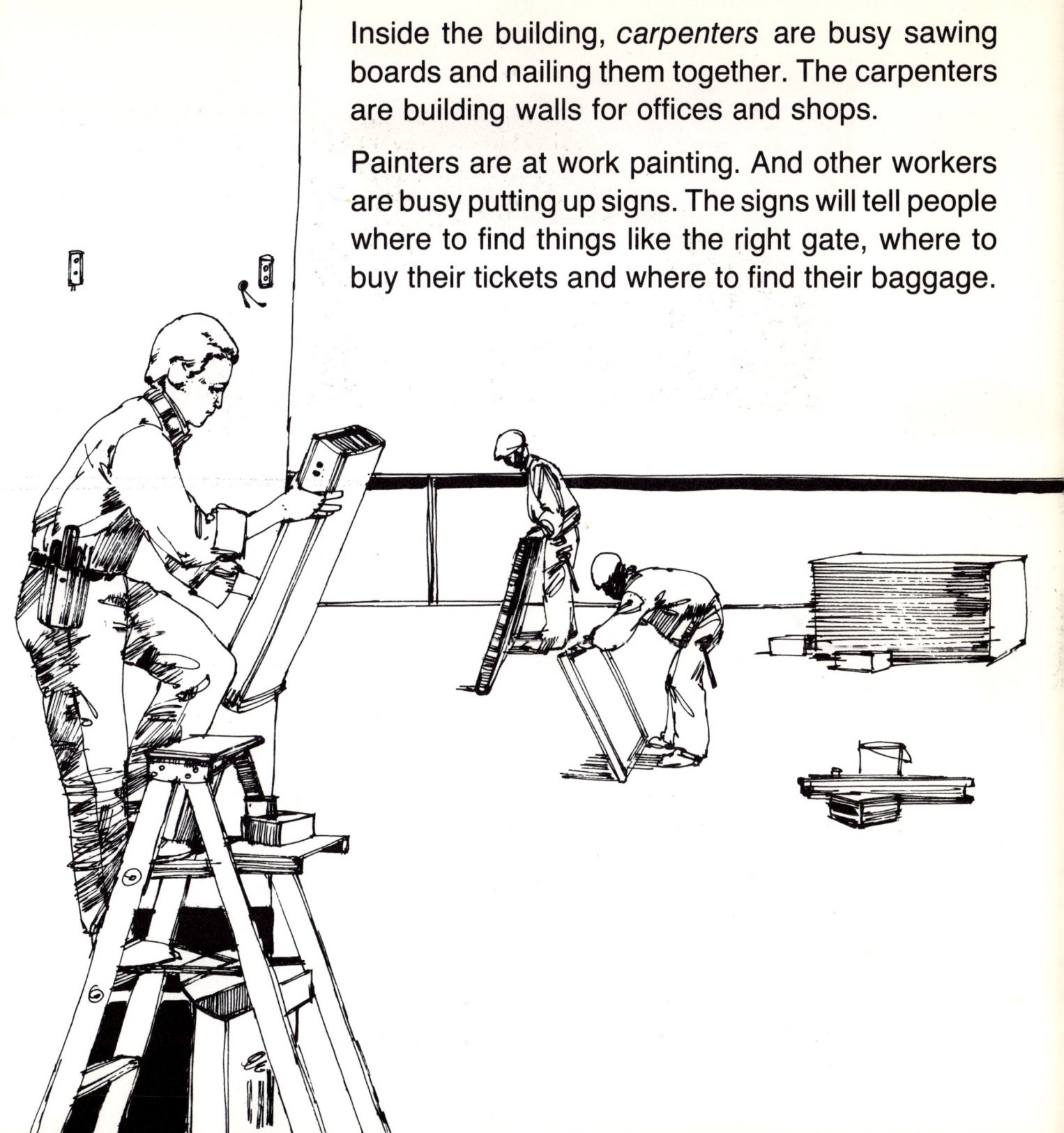

HANGARS AND OTHER BUILDINGS

While the terminal building is being built, other buildings are going up. One funny looking building is called a *maintenance hangar*. This building is very wide. Its doors are so big that one whole side of the building opens up. It must be wide because airplanes are put in this building. Inside mechanics work on airplanes.

Airplanes carry letters and packages, called *cargo*. So a building for cargo must also be built. It has a *dock* on each side. A dock is like a long porch. It is high enough to load and unload cargo from the airplanes on one side and trucks on the other.

TAXIWAYS AND APRONS

If you fly over an airport and look down at it, you can see roadways crossing the runways. These are *taxiways*. Airplanes use them when they move, or taxi, between the runways and the aprons.

BUILDING THE AIRFIELD

The runways, taxiways and aprons are all part of the *airfield,* which is very flat. The airfield is made flat by big machines which cut down the high places and fill in the low places.

Most of the dirt is moved by machines called *motor scrapers.* The motor scraper cuts a layer of dirt from the high places, then carries it to a low place, and dumps it.

When they are loading, scrapers sometimes need to be pushed by *crawler tractors*. Crawler tractors are also used to push dirt around. When they do this, they use wide steel blades in front, and are called *bulldozers*.

Next, the dirt must be smoothed by a machine called a *motor grader*.

The ground must be firm, too. It must be able to support the airplanes that will soon run over it.

Heavy rollers, called *compactors,* are moved back and forth over the dirt like rolling pins. They pack the earth down until it is hard.

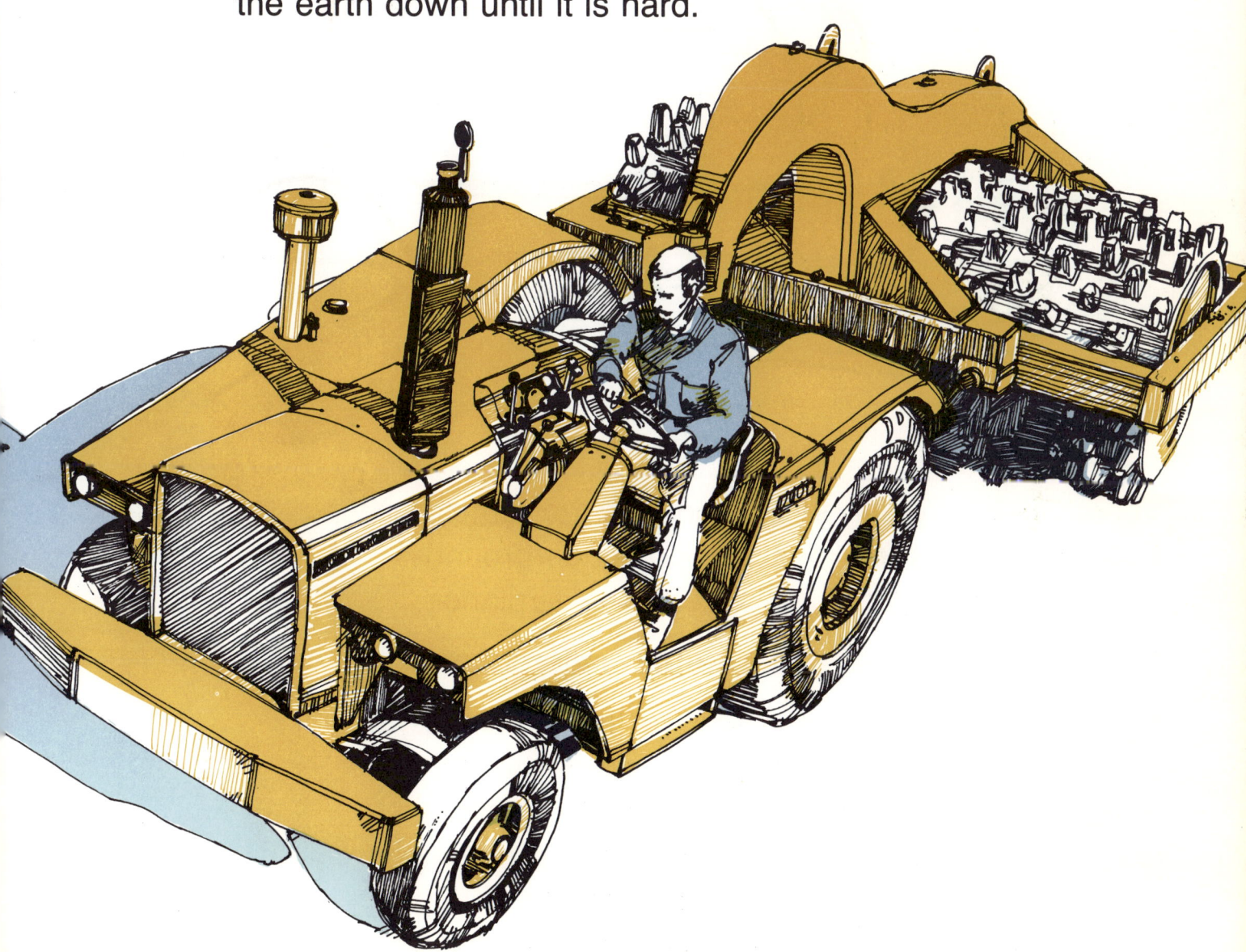

UNDERGROUND UTILITIES

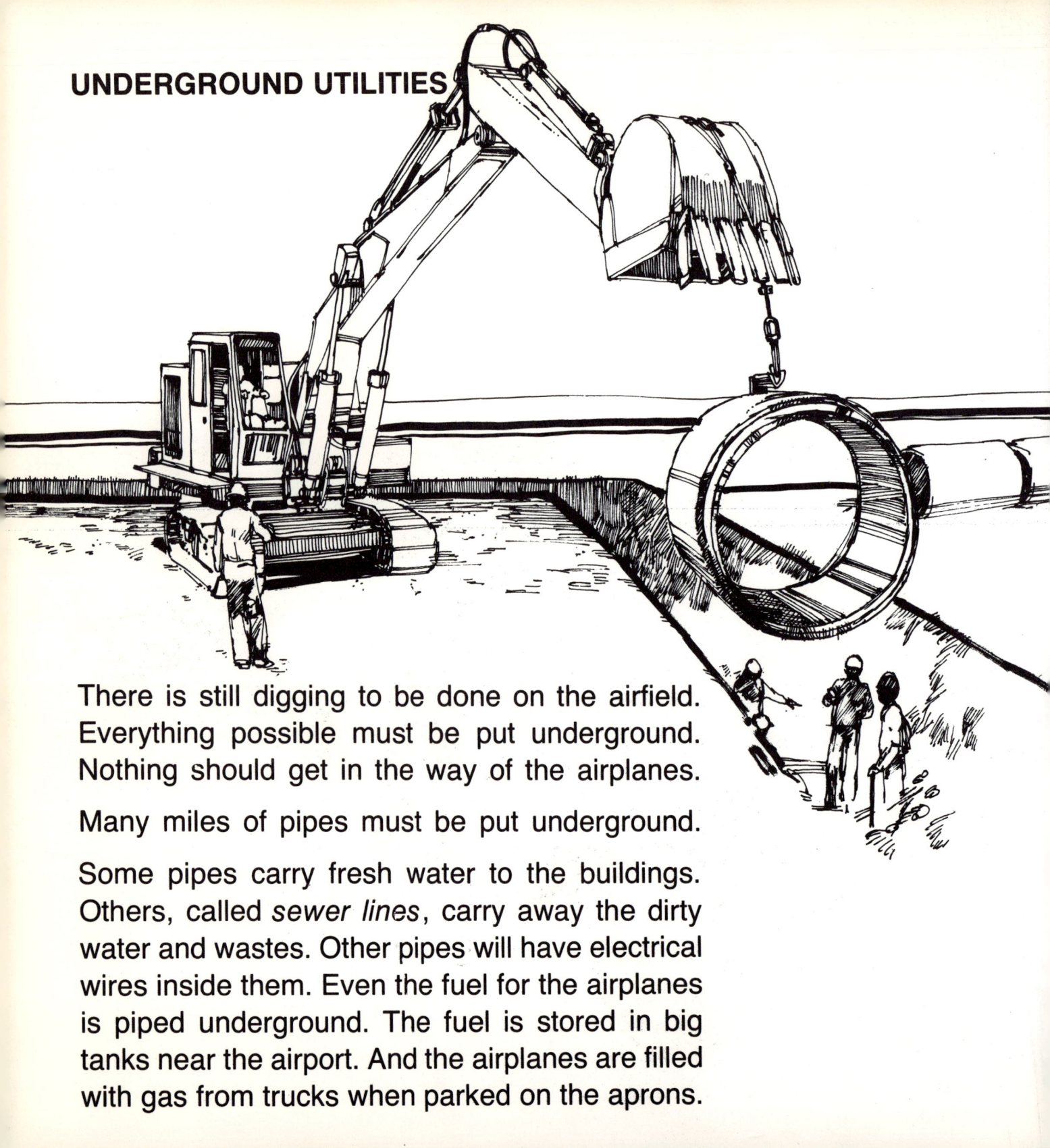

There is still digging to be done on the airfield. Everything possible must be put underground. Nothing should get in the way of the airplanes.

Many miles of pipes must be put underground.

Some pipes carry fresh water to the buildings. Others, called *sewer lines*, carry away the dirty water and wastes. Other pipes will have electrical wires inside them. Even the fuel for the airplanes is piped underground. The fuel is stored in big tanks near the airport. And the airplanes are filled with gas from trucks when parked on the aprons.

The biggest underground pipes will drain the water from the airfield when it rains. The *drainage pipes* are often so big that you could stand up inside them.

The water runs off the airfield into the drainage pipes through *surface drains*. The drains have steel cages over them, so an airplane wheel can run right over them and not fall in.

One machine that digs ditches for pipes is called a *backhoe*. It has a big heavy steel arm. On the end of this arm is a large steel bucket turned upside down. When the machine digs, it looks like a big animal digging with one paw.

A faster digger is the *ditching machine*. Some ditching machines have a big wheel on the front. The wheel turns backwards and has teeth on it that bite into the dirt.

After the pipes are laid in the ditch, a bulldozer pushes the dirt back in the ditch.

PAVING THE AIRFIELD

Now the runways, taxiways and aprons are ready to be paved. First, a big machine called a *fine grader* smooths the ground where the pavement will go.

Crushed rock is then spread evenly by a *rock spreader*. And it is packed down by a compactor.

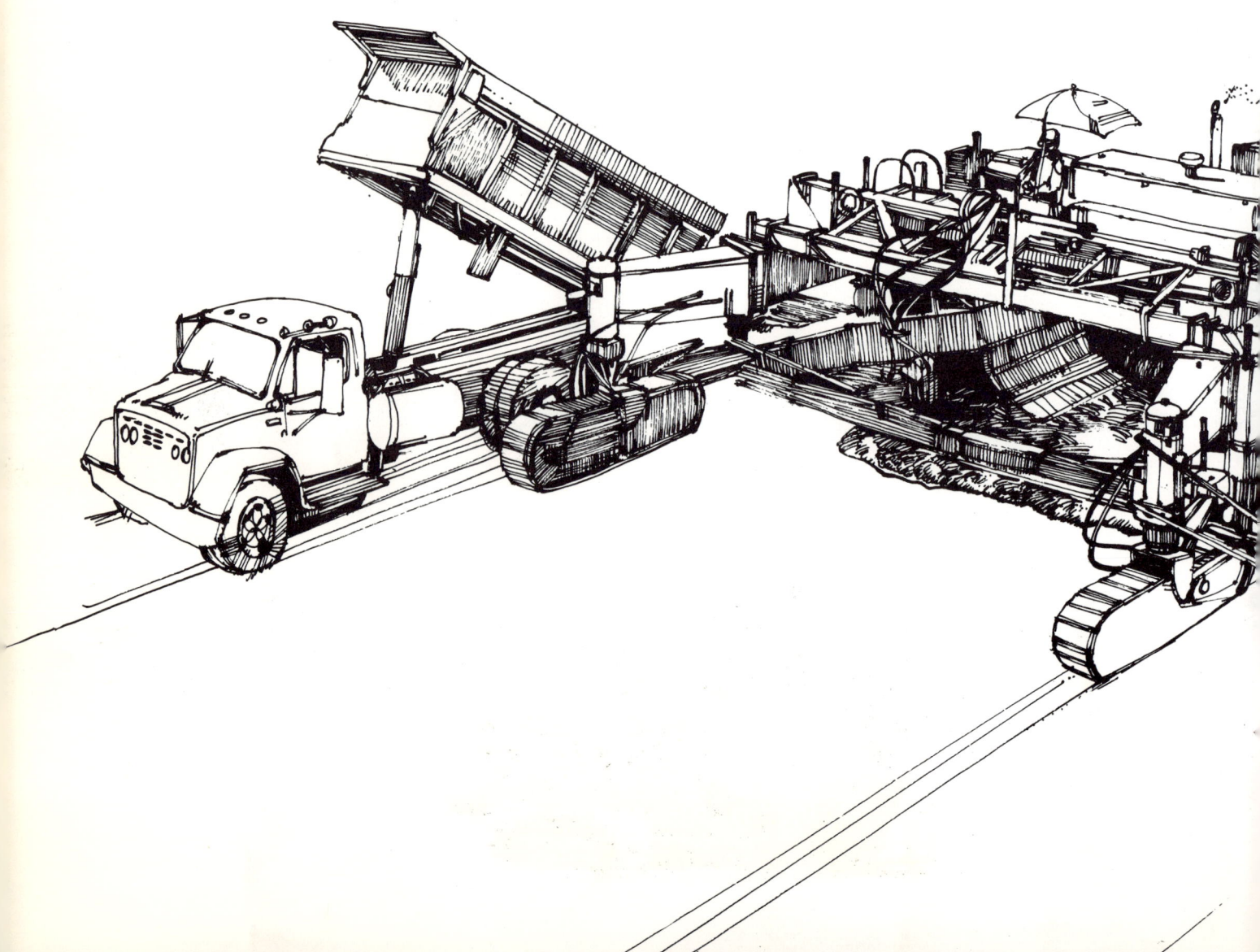

Everything is now ready for the paving crew. This is a very exciting time, as the *paving train* slowly comes clanking and grumbling along. The paving train has several machines in a row, one after the other.

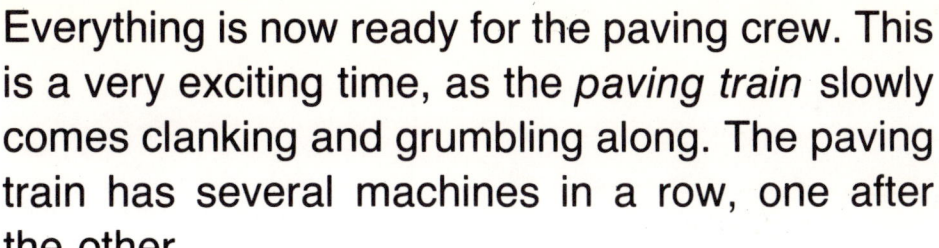

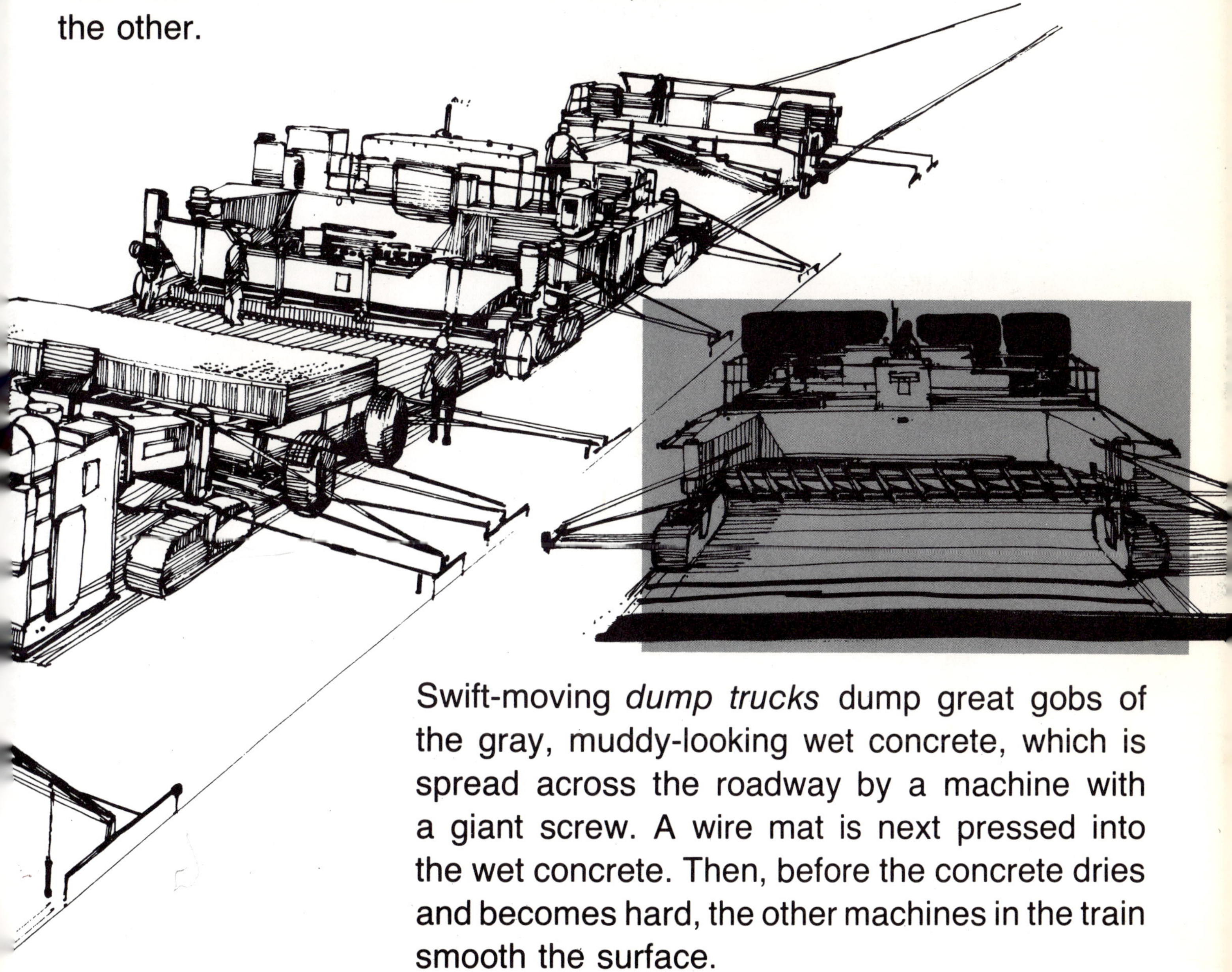

Swift-moving *dump trucks* dump great gobs of the gray, muddy-looking wet concrete, which is spread across the roadway by a machine with a giant screw. A wire mat is next pressed into the wet concrete. Then, before the concrete dries and becomes hard, the other machines in the train smooth the surface.

The concrete pavement on an airfield is many times thicker than the concrete in a sidewalk or driveway. Places where the airplanes will stand still, such as the parking aprons and the ends of the runways, are called *critical areas*. The pavement must be thickest of all in the critical areas.

Trucks bring the wet concrete from a nearby *central mix plant*. This is a small factory where piles of cement, sand, and crushed stone are mixed with water to form the wet concrete.

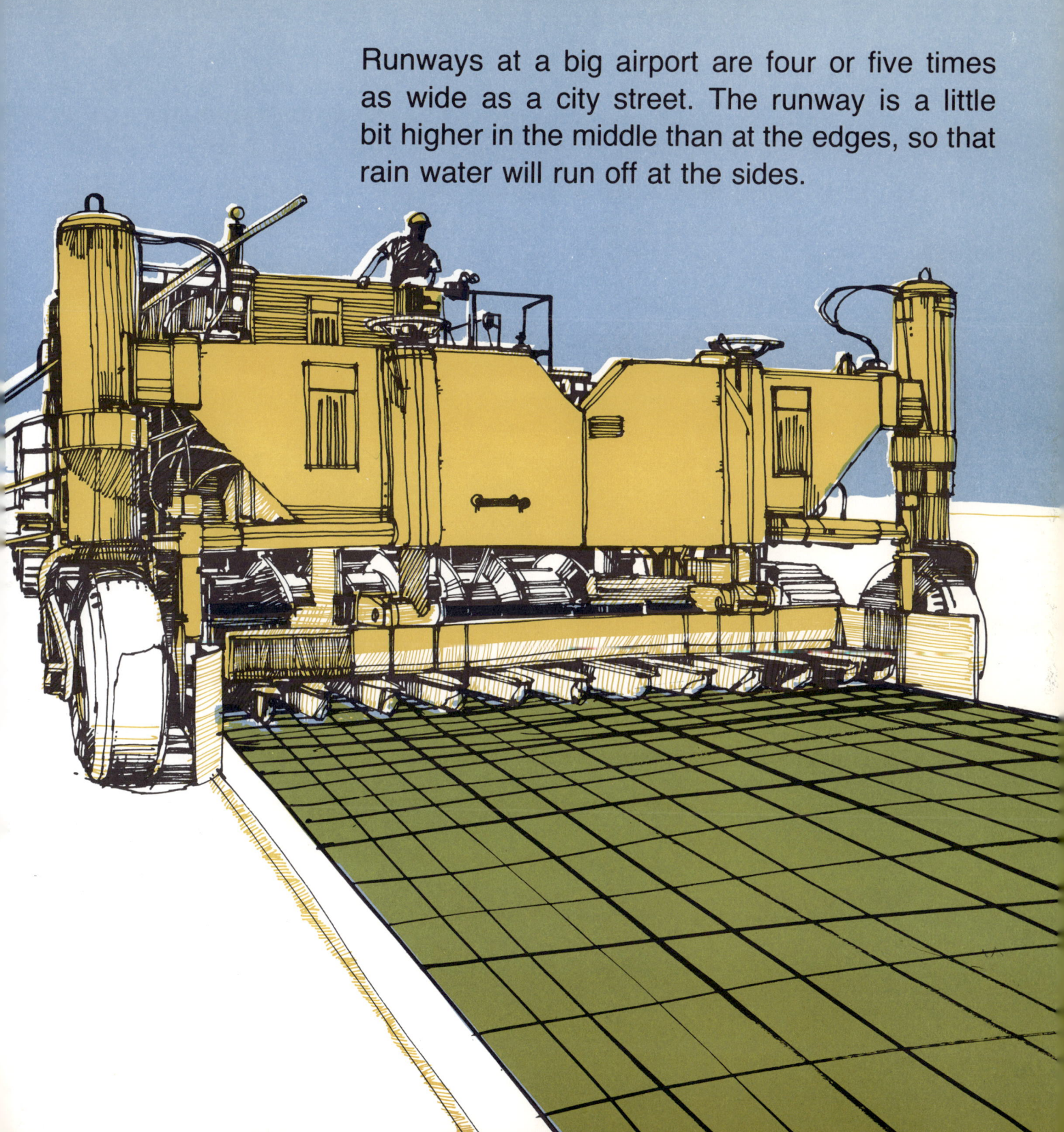

Runways at a big airport are four or five times as wide as a city street. The runway is a little bit higher in the middle than at the edges, so that rain water will run off at the sides.

On each side of the runways and taxiways is a strip of black asphalt paving. This is called a *shoulder*. The shoulder helps to keep dust from blowing when the airplanes go by. Some of the big airplanes have jet engines out near the end of their wings. These engines hang out past the edge of the runway, and would stir up huge clouds of dust if there were no shoulders.

AIRPORT LIGHTING

Even after the paving is finished there is still much work to be done on the airfield.

Different colored lights must be placed on the airfield. At night these lights will make the airfield look like a big Christmas tree. But they are not decorations. At night the lights show the pilots where to land and taxi their airplanes.

Electricians place a row of green lights across the end of each runway. These are called *threshold lights*. They show the pilots where the ends of the runways are.

A row of clear lights is put on each side of the runway. These lights are called *runway edge lights*.

The only lights that are right on the runway are called *center line lights*. These lights do not stick up in the air. They are put in holes drilled in the concrete.

Taxiways have edge lights, too. They are blue.

Anything on the ground that sticks up into the air is called an *obstruction*. Buildings, trees, high wires and mountains near an airport are obstructions. And if they cannot be torn down, they must be marked by *obstruction lights*. These are red. And the pilots know that these lights mean BEWARE!

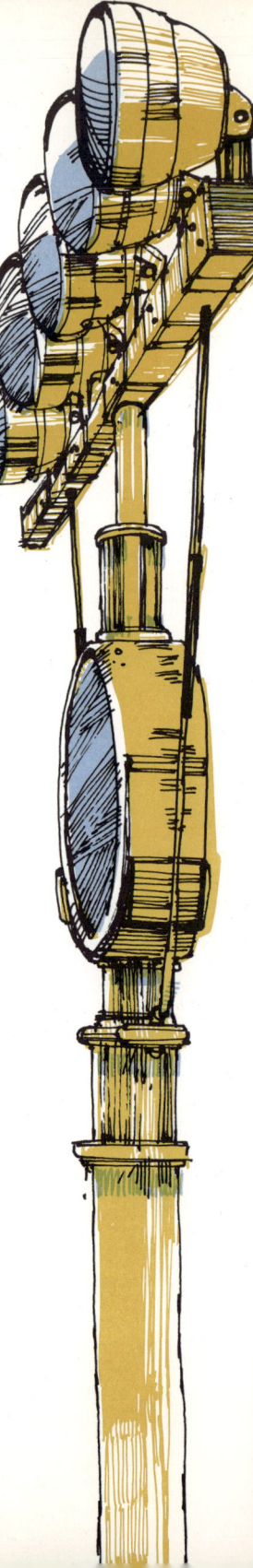

Usually the first light that you see when you come near an airport is the *beacon*. This light is high above the airfield. The beacon light shines in two directions at once, and turns around and around. Sometimes the beacon light is on top of the control tower.

All of the lights at the airfield are controlled from the air traffic control tower.

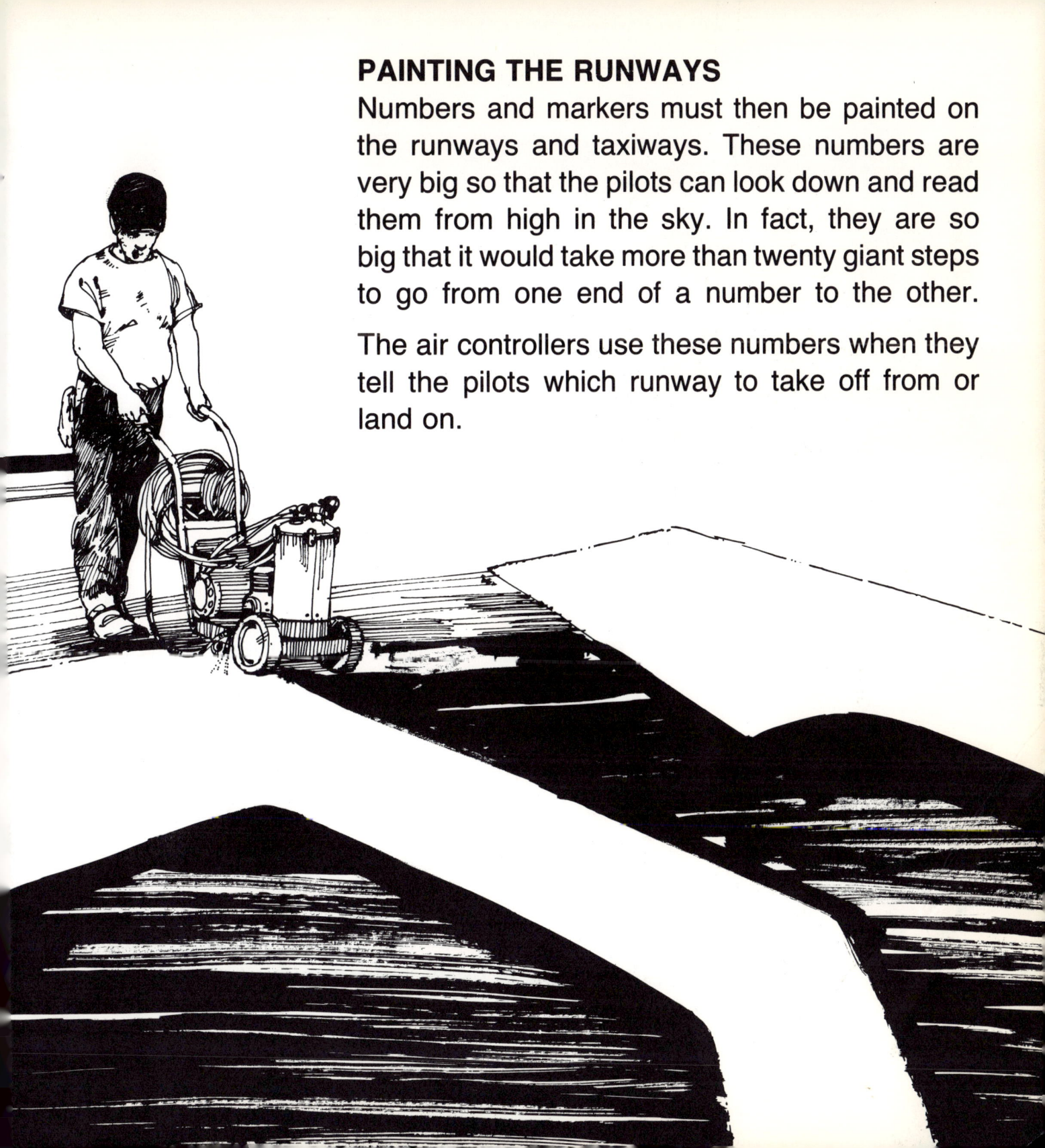

PAINTING THE RUNWAYS

Numbers and markers must then be painted on the runways and taxiways. These numbers are very big so that the pilots can look down and read them from high in the sky. In fact, they are so big that it would take more than twenty giant steps to go from one end of a number to the other.

The air controllers use these numbers when they tell the pilots which runway to take off from or land on.

Some runways have *aiming marks*. The pilots point their big airplanes at these marks when they are landing. These are bigger than the runway numbers. It takes more paint to cover one of these aiming markers than to paint a large house.

There is so much painting to do that a special machine is used.

One job the painting machine is used for is to paint wide lines down the middle of runways and taxiways. White paint is used on the runways. Yellow paint is used on the taxiways.

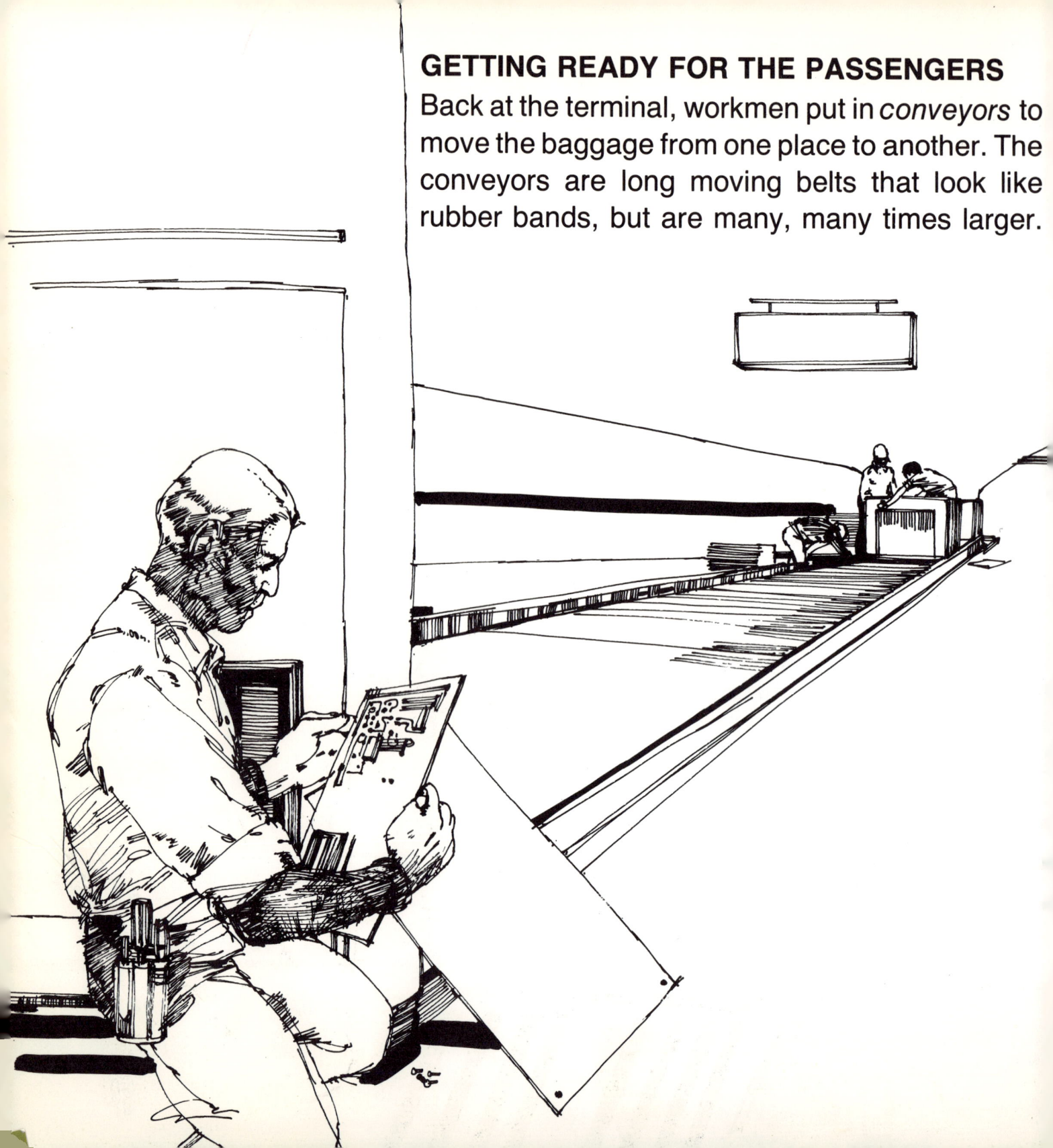

GETTING READY FOR THE PASSENGERS

Back at the terminal, workmen put in *conveyors* to move the baggage from one place to another. The conveyors are long moving belts that look like rubber bands, but are many, many times larger.

Other workmen attach *jetways* to the gates. These are passageways which can be stretched out between the gate and the airplane. Passengers can walk through a jetway to the airplane without going outside.

The airport will soon open. Workmen are busy putting in lights in the building and outside. Other men plant shrubs and trees to make the airport beautiful.

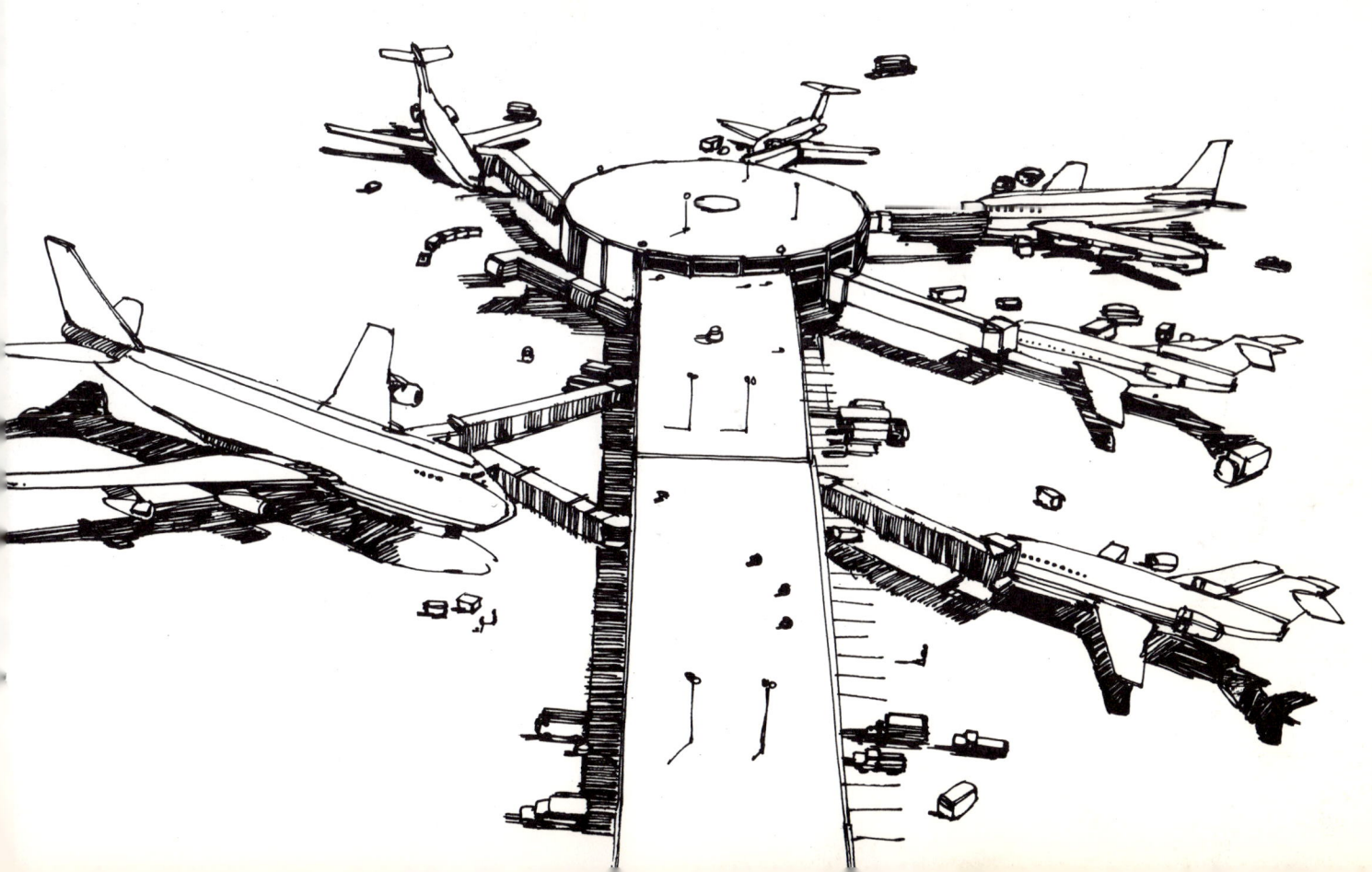

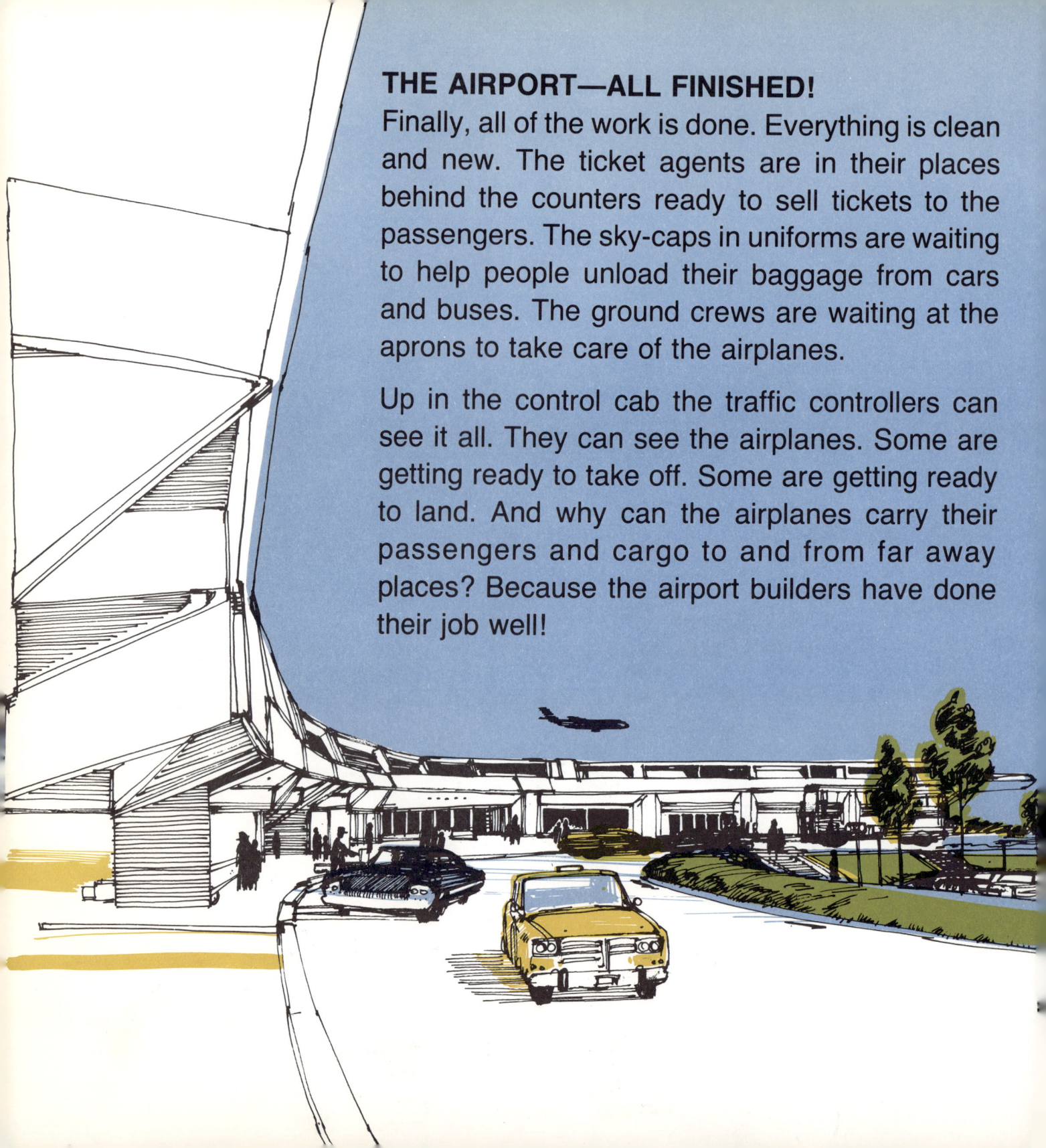

THE AIRPORT—ALL FINISHED!

Finally, all of the work is done. Everything is clean and new. The ticket agents are in their places behind the counters ready to sell tickets to the passengers. The sky-caps in uniforms are waiting to help people unload their baggage from cars and buses. The ground crews are waiting at the aprons to take care of the airplanes.

Up in the control cab the traffic controllers can see it all. They can see the airplanes. Some are getting ready to take off. Some are getting ready to land. And why can the airplanes carry their passengers and cargo to and from far away places? Because the airport builders have done their job well!

About the Authors

The authors of THE AIRPORT BUILDERS are two very active men in the construction industry. And with the help of their wives—both experienced teachers—their expertise was directed to the young reader.

James E. Kelly is the educational director of the Heavy Constructors Association of Greater Kansas City, a chapter of the Associated General Contractors of America.

Besides following major national construction projects, he also has an interest in government, currently serving as president of the Kansas City International Relations Council to study foreign affairs; he was formerly mayor pro-tem of Independence, Missouri, where he and his wife now live.

William R. Park is now principal construction economist at Midwest Research Institute. He is a licensed professional engineer, active in the American Society of Civil Engineers and other technical and professional organizations.

He is the author of two technical books and has written more than a hundred articles for a variety of construction journals. Mr. Park and his family live in Prairie Village, Kansas.

About the Artist

Joel Snyder has illustrated several children's books since graduating from the Rhode Island School of Design.

Mr. Snyder, his wife and twin boys live in Farmingville, New York where they enjoy camping and golfing.

```
629.136   Kelly, James E
Kel
         The airport builders
```

DATE DUE			

SCHOOL DISTRICT NO. 48
Beaverton, Oregon